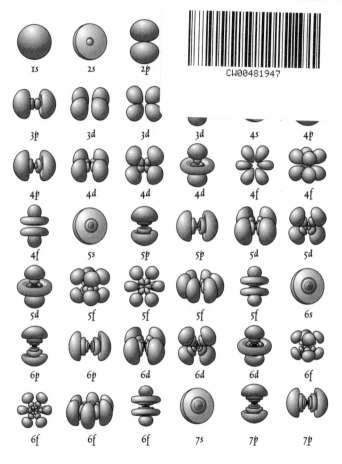

Electron Orbitals

First published 2023
This edition © Wooden Books Ltd 2023

Published by Wooden Books Ltd.
Glastonbury, Somerset
www.woodenbooks.com

British Library Cataloguing in Publication Data
Tweed, M.
The Elements of Chemistry

A CIP catalogue record for this book
may be obtained from the British Library

ISBN-10: 1-907155-52-x
ISBN-13: 978-1-907155-52-9

Designed and typeset in Glastonbury, UK.

Printed in China on FSC® certified papers by
RR Donnelley Asia Printing Solutions Ltd.

THE ELEMENTS OF
CHEMISTRY

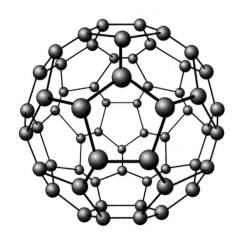

Matt Tweed

This book was made possible by many great minds, past, present & future. Love to C for putting up with me & J for getting me to look again...

Readers might find these titles of interest: *Molecules: The Elements and the Architecture of Everything*, Theodore Gray; *How to Build a Habitable Planet*, Charles H. Langmuir & Wallace S. Broecker; *The Story of Earth*, Robert M. Hazen; *Trespassing on Einstein's Lawn*, Amanda Gefter; *QBism: The Future of Quantum Physics*, Hans Christian von Baeyer; *Reality Is Not What It Seems*, Carlo Rovelli; *The God Equation*, Michio Kaku; *Wholeness and the Implicate Order*, David Bohm; *A Journey into Gravity and Spacetime*, John Archibald Wheeler.

INGREDIENTS (*makes one human*): *oxygen 61%, carbon 23%, hydrogen 10%, nitrogen 2.6%, calcium 1.4%, phosphorus 1.1%, potassium 0.2,%, sulfur 0.2%, sodium 0.1%, chlorine 0.1%, plus magnesium, iron, fluorine, zinc, and other trace elements.*

Contents

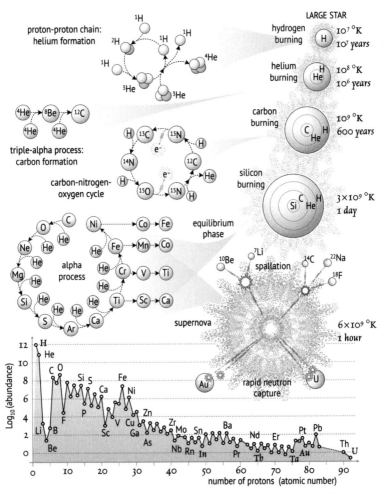

NUCLEOSYNTHESIS: elements up to iron with an even number of protons are created through the alpha process. Odd numbered elements form by less efficient slow neutron capture, resulting in a prominent zig-zag in their relative cosmic abundance.

INTRODUCTION

CHEMISTRY BEGAN with a big bang, 13.8 billion years ago. As the pinpoint of intense energy expanded and cooled, first photons, then subatomic quarks and leptons condensed out of the seething quantum vacuum. One second later, quarks merged into protons and neutrons, which minutes later combined into nuclei of hydrogen (75%), helium (25%), and traces of lithium. Some 377,000 years later, the universe had cooled enough for electrons to settle into orbits around these nuclei, forming atoms. After another 200 million years, gravity had clumped these gases together tightly enough to ignite the first stars.

Inside stars, for most of their life, hydrogen atoms fuse into helium, releasing more energy. When this fuel runs out they collapse under gravity, their soaring temperatures and pressures melding hydrogen and helium into carbon, nitrogen, and oxygen. In massive stars, this process repeats, as carbon is crunched into oxygen, neon, and magnesium, while slow neutron capture assembles intermediate elements. Further cycles produce silicon and sulfur, before, in a last collapse, the formation of cobalt, nickel, and iron stall further fusion. Gravity once again takes over, with a terminal implosion followed by a blazing supernova explosion, in which new elements are created either by rapid neutron capture, with nuclei colliding to create heavy elements like uranium, or spallation, where large atoms are smashed into smaller pieces.

Three generations of stars have now recycled the primordial hydrogen and helium into a palette of just over a hundred types of atom. Together, these combine to form the astonishing mosaic of our visible universe.

MATTER MAGICKS
alchemy transmutes into chemistry

The investigation of matter stretches back into the distant past, to our ancestors painting with colour on cave walls, learning the secrets of fire, and heating particular rocks to obtain wondrous metals (*opposite*).

The ancient Egyptians knew of seven metals, and carbon and sulfur. Their art of *khemia* linked the metals to the planets, assigning them unique qualities (*below, left*). In India, sages spoke of the three *gunas*, fire, earth, and water, while the Chinese system of *Wu-hsing* recognised two more substances, metal and wood (*below, mid*). For the Greeks, all things originated from earth, air, fire, and water (*below, right*). Aristotle [384–322 BC] named these four principles *elements*, adding a fifth, *quintessence*, to form the heavens. The philosopher Democritus [460–370 BC] proposed that dividing matter *ad infinitum* would eventually leave an indivisible *atmos*, an idea that was subsequently forgotten for centuries.

The quest spread to medieval Arabia with the study of *Al-khemia*, thence crossing to Europe, where gold-seeking *alchemists* searched for the *Gloria Mundi* or philosopher's stone. Slowly, through experimentation, intuition, and plenty of accidents, the science of *chemistry* was born.

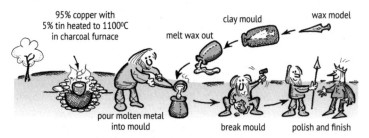

ABOVE: **BRONZE**, *an alloy of copper and tin, is stronger than its constituents. First worked around the 4th millennium BC, it caused a technological revolution. The metals are extracted by smelting — heating the ores in a fire and separating the molten runoff. Above, a spear head is cast using the lost wax process.*

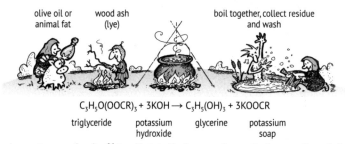

$$C_3H_5O(OOCR)_3 + 3KOH \rightarrow C_3H_5(OH)_3 + 3KOOCR$$

triglyceride   potassium   glycerine   potassium
hydroxide   soap

ABOVE: **SOAPS** *are the salts of fatty acids. Saponification occurs when an oil or fat reacts with an alkali to produce an alcohol and a soap-salt that has both hydrophilic (water-attracting) and hydrophobic (water-repelling) properties. Perhaps discovered when fat dripped into damp ashes in ancient Babylon.*

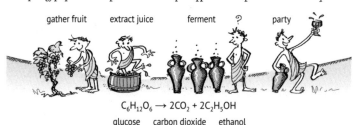

$$C_6H_{12}O_6 \rightarrow 2CO_2 + 2C_2H_5OH$$

glucose   carbon dioxide   ethanol

ABOVE: **FERMENTATION** *uses yeasts or bacteria to metabolically break down sugar into carbon dioxide and alcohol. Humanity has been enjoying the results for at least thirteen millennia.*

PHASES & FORMS
jewels in crystal

Experiments by John Dalton [1766–1844] comparing the weights, masses and density of substances, revealed that many assumed elements were in fact MOLECULES, compounds made of different smaller ATOMS.

Matter can exist in several different states or PHASES (*opposite top*):

SOLID At low temperatures and high pressures, atoms and molecules pack closely together. *Crystals* are regular three-dimensional lattices that balance between attractive and repulsive forces (*below and opposite lower*).

LIQUID Adding energy by heating vibrates the ordered structures apart, allowing the atoms and molecules to flow and change shape.

GAS Heating more further weakens the molecular bonds, letting the molecules and atoms scatter in all directions to fill a container.

PLASMA At even greater temperatures, all molecular bonds are broken and even the atoms are shaken apart to form an electrically charged, *plasma*, such as that found in the superheated corona of the Sun.

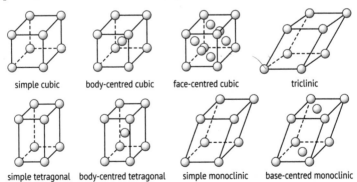

simple cubic	body-centred cubic	face-centred cubic	triclinic
simple tetragonal	body-centred tetragonal	simple monoclinic	base-centred monoclinic

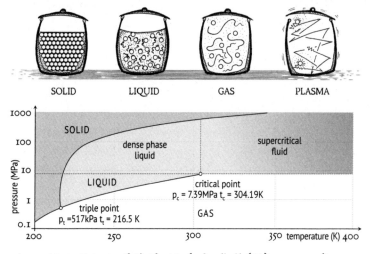

SOLID   LIQUID   GAS   PLASMA

ABOVE: A PHASE DIAGRAM, plotting the state of carbon dioxide (CO_2) over a range of pressures and temperatures. Solid, liquid and gas phases coexist at the TRIPLE POINT, below which the solid sublimes into a gas without becoming a liquid. At the CRITICAL POINT, gas and liquid phases become indistinguishable, transforming into a supercritical fluid with properties of both.

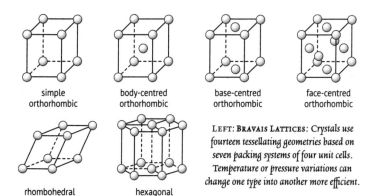

simple orthorhombic

body-centred orthorhombic

base-centred orthorhombic

face-centred orthorhombic

rhombohedral

hexagonal

LEFT: BRAVAIS LATTICES: Crystals use fourteen tessellating geometries based on seven packing systems of four unit cells. Temperature or pressure variations can change one type into another more efficient.

INSIDE THE ATOM
protons, neutrons, and electrons

Atoms consist of a tiny central NUCLEUS orbited by one or more ELECTRONS (*opposite top*). The nucleus, around one to ten femtometers ($10^{-15}m$) across, contains two types of particle: PROTONS and NEUTRONS (the NUCLEONS). Protons carry a single positive electric charge, while neutrons, although of similar size and mass, have no electric charge.

The number of protons in a nucleus (its ATOMIC NUMBER) defines an element and determines its overall chemical behavior. However the number of neutrons can vary, giving a family of ISOTOPES that react the same chemically but differ in mass and other properties, such as nuclear stability and radioactive decay rates (*see p. 42*).

Surrounding the nucleus is a cloud of negatively charged electrons, one for each proton (*opposite top*). With a mass almost two thousand times less than a proton, the negatively charged electrons are strongly attracted to the positively charged protons, yet repel each other, ignoring the chargeless neutrons. To navigate the forces, the electrons dance around the nucleus in ORBITALS, three dimensional patterns that become progressively complex as atoms get larger. Each orbital can contain one electron pair, forming SHELLS of increasing energy (*opposite lower*).

Amazingly, atoms are almost entirely empty space. The mass of an electron orbiting a nucleus is equivalent to a cat swinging a bee on the end of a half-mile long piece of elastic.

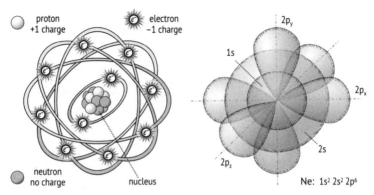

proton
+1 charge

electron
−1 charge

neutron
no charge

nucleus

2p_y

1s

2p_x

2s

2p_z

Ne: $1s^2\ 2s^2\ 2p^6$

ABOVE: The CLASSICAL representation of a neon atom: a central nucleus of 10 protons and 10 neutrons surrounded by 10 electrons, two in an inner orbit, the rest in outer ones.

ABOVE: Simplified QUANTUM MECHANICAL view of a neon atom: the shape of an orbital lobe represents the probability of finding an electron as given by the Schrödinger wave equation.

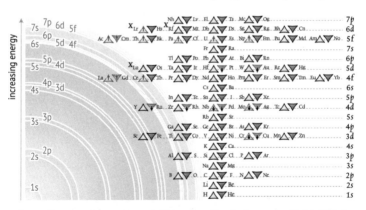

ABOVE: ELECTRON ORBITAL SHELLS build in sequence from the innermost 1s. Each half fills △ with electrons before completing ▽ as oppositely spinning pairs. A few exceptions break the rule, either by borrowing an electron from the s-orbital below ⏶ or jumping up to the orbital above ⏶. An X indicates d-orbitals that begin to fill before the electrons drop to the f-orbital below.

PERIODIC TABLES
elemental arrangements

The most common method of ordering atoms is the **PERIODIC TABLE** of elements (*opposite top*), first formulated by Dmitri Mendeleev in 1869 based on the way chemical properties fall into recurring patterns. Modern versions of his table are arranged by increasing atomic number, placing vertical **GROUPS** of atoms with the same pattern of outer electrons (and hence behaviour) alongside horizontal **PERIODS** of electron orbital sets. **BLOCKS** define the different electron orbitals.

By contrast, the **STOWE TABLE** (*lower opposite*) uses each element's unique descriptive series of *quantum numbers* to show the ordering of the intricate orbital sets of electron shells, with the innermost at the top. Another alternative layout, **BENFEY'S SPIRAL** (*below*), emphasizes the continuity of the elements, with groups radiating out from a hydrogen hub and blocks forming outcrops. It even shows a conjectured, but as yet undiscovered, g-block of super-heavy elements.

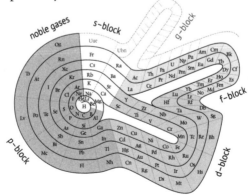

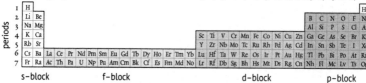

ABOVE: THE PERIODIC TABLE is read left to right, top to bottom, ignoring the spaces, with increasing atomic number. Horizontal periods show the filling of electron shells. Vertical groups (numbered) have similar outer electrons. Blocks delineate the electron orbitals (see also pages 56-57).

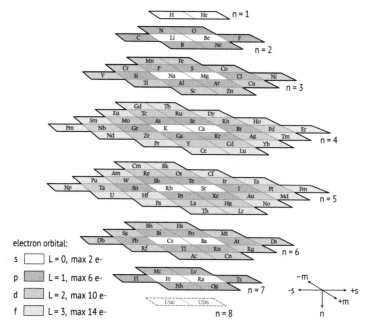

ABOVE: THE STOWE TABLE is ordered using an element's unique series of quantum numbers. It shows the successive electron shells (n) and the orbital sets of which they are comprised. Each successive orbital (L) contains one more pair (m) of paired electrons of opposite spin (s).

ORBITAL STRUCTURES
the whirly world of the very small

At the scale of fundamental particles like electrons, energy comes in discrete packets, or QUANTA. Matter behaves as both particles *and/or* waves, depending on what is being measured (*opposite top right*).

Mathematical WAVE FUNCTIONS calculate the probability of finding an electron in a specific place. The first *1s*-orbital shell is spherical with electrons that may pass by the nucleus. The second, the *2p*, builds into three SETS of double teardrops reflecting around NODAL PLANES where electrons won't go (*below*). Electrons are mostly found within their main density plot, although there is a nonzero chance that they could be anywhere (*opposite top left*). Pumping more energy into an atom causes excited electrons to jump into new higher orbitals, the rising energy forcing the electrons into increasingly exotic journeys, filling flowering SUB-SHELLS. The cycle repeats and orbitals complete, replete with twin electrons of the opposite quantum SPIN (*opposite, p 7. and frontispiece*).

The wave/particle duality of electrons renders them subject to the UNCERTAINTY PRINCIPLE, making it impossible to know complementary values like position *and* momentum simultaneously. In addition, the very act of observing something so tiny can impact its behaviour and adversely affect any measurement.

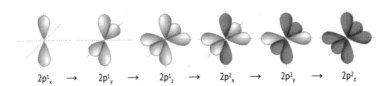

$2p^1_x$ $\rightarrow$ $2p^1_y$ $\rightarrow$ $2p^1_z$ $\rightarrow$ $2p^2_x$ $\rightarrow$ $2p^2_y$ $\rightarrow$ $2p^2_z$

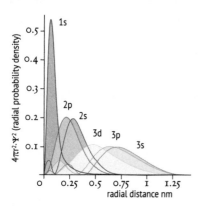

ABOVE: ELECTRON ORBITAL RADII:
The probability of finding an electron in a particular place is related to the square of its wave function (Ψ).

ABOVE: MATTER WAVES: UPPER: *The first three orbitals as standing waves.* **LOWER:** *The envelope of a moving electron's wave packet.*

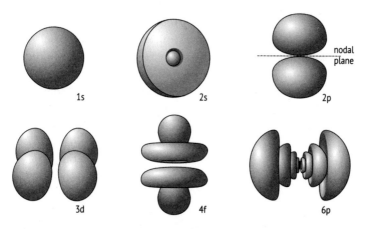

ABOVE: Idealized **ELECTRON ORBITALS** of a hydrogen atom. With increasing energy, the single electron jumps to higher orbitals. With greater atomic mass, electron interactions and other quantum effects make the interweave around heavier atoms ever more complex.

BONDING
an attractive proposition

Elements bond to form compounds by rearranging only their outer, highest energy VALENCE electrons. Losing or gaining electrons causes an atom to become an electrically charged ION. Metals are ELECTROPOSITIVE, losing electrons to create positively charged CATIONS. Nonmetals are ELECTRONEGATIVE, gaining electrons to make negatively charged ANIONS.

An IONIC BOND forms when a cation lends electrons to an anion, giving both stable full outer orbitals (*opposite top left*). Tough and brittle with high melting points, many ionic compounds dissolve in water.

Non-metals combine via COVALENT BONDS, which shuffle and share outer electrons into pairs to fill any empty orbitals (*opposite top right*). The attraction of the electrons for the positively charged nuclei outweighs any mutual repulsion, holding the resulting molecule together.

Hydrogen atoms attached to non-metals push against any unbonded electrons, creating a tiny charge difference across the molecule. If another electronegative atom is nearby, a weak HYDROGEN BOND forms between the slightly positive hydrogen atoms of one molecule and the negative unbonded *lone pair* electrons of the other (*opposite mid left*).

With overlapping orbitals forming additional π-BONDS (*opposite mid right*), and asymmetrical electrons causing instantaneous small VAN DER WAALS FORCES (*below*), molecular glue comes in many varieties.

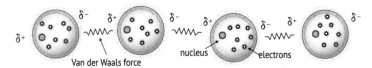

Van der Waals force

nucleus   electrons

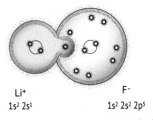

Li⁺
$1s^2 2s^1$

F⁻
$1s^2 2s^2 2p^5$

IONIC BONDING: In lithium fluoride (LiF), the lithium cation 'lends' its single valence electron to the fluorine anion, allowing it to completely fill its outer electron shell.

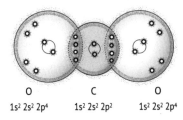

O
$1s^2 2s^2 2p^4$

C
$1s^2 2s^2 2p^2$

O
$1s^2 2s^2 2p^4$

COVALENT BONDING: In the carbon dioxide molecule (CO_2), the valence electron pairs are shared between the atoms giving each a stable full outer shell.

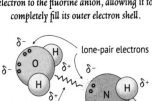

A **HYDROGEN BOND** formed between water (H_2O) and ammonia (NH_3). The small charge difference (δ^+, δ^-) across the molecules weakly attracts the positive and negative ends.

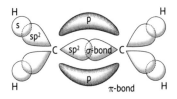

The carbon **DOUBLE BOND** in ethylene ($H_2C = CH_2$). The covalent σ-bond between two sp^2-orbitals is reinforced by a slightly weaker π-bond, due to overlapping p−orbitals.

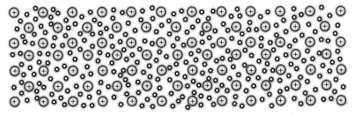

METALLIC BONDING: In metals, electrons escape from their nuclei and dissociate into a 'sea' around a lattice of positive ions. The conductivity and shininess of metals is due to these mobile electrons, with their strength and high melting points resulting from the attraction between ions and electrons.

REACTIONS
chemical conflagrations

Chemical **REACTIONS** transform one substance into another. The process can be shown as an equation, with the initial **REACTANTS** on the left and final **PRODUCTS** to the right of an arrow ($\rightarrow$) if the reaction is one way, or a double arrow ($\rightleftharpoons$) for a two-way equilibrium. Phase can be indicated by (*s*) for solid, (*l*) for liquid, (*g*) for gas, or (*aq*) for an aqueous solution. Equations must balance, giving equal quantities of atoms on both sides.

Reactions can be categorised into four main types:

SYNTHESIS: Two or more compounds are combined into a product;

$$A + B \rightarrow AB \qquad e.g. \qquad C + O_2 \rightarrow CO_2$$

DECOMPOSITION: Breaks down a compound into simpler products;

$$AB \rightarrow A + B \qquad e.g. \qquad 2AgBr \rightarrow 2Ag + Br_2$$

SINGLE DISPLACEMENT: One element is replaced by another;

$$A + BC \rightarrow AC + B \qquad e.g. \qquad K + NaCl \rightarrow KCl + Na$$

DOUBLE DISPLACEMENT: Elements swap to form new compounds;

$$AB + CD \rightarrow AC + BD \qquad e.g. \qquad Na_2S + 2HCl \rightarrow 2NaCl + H_2S$$

EXOTHERMIC reactions release energy, such as heat, to their surroundings. An explosive example is when hydrogen and oxygen meet a flame and the atoms shuffle electrons to form water, releasing heat and light (*opposite top*). Conversely, **ENDOTHERMIC** reactions absorb energy, such as plants photosynthesizing the sun's energy (*lower opposite*).

Laboratory chemicals are often measured in **MOLES** (*mol*), where one mole contains a fixed number, **AVOGADRO'S CONSTANT** ($6.02214076 \times 10^{23}$), of atoms or molecules. For example, the number of atoms in 1 gram of carbon = Avogadro's number divided by the atomic weight of carbon.

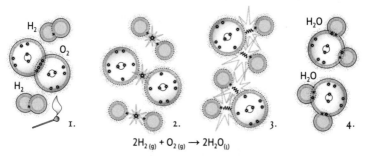

$$2H_{2\,(g)} + O_{2\,(g)} \rightarrow 2H_2O_{(l)}$$

AN EXOTHERMIC REACTION: 1. Adding heat to hydrogen (H_2) and oxygen gas (O_2) provides energy to 2. break the bonds forming the molecules. 3. The atoms rearrange into a more efficient configuration, releasing energy as even more heat. 4. Covalently bonded water molecules (H_2O).

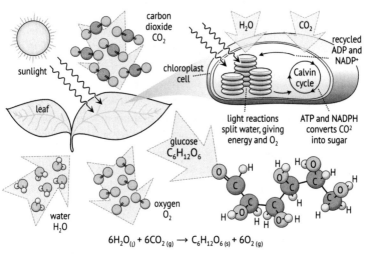

$$6H_2O_{(l)} + 6CO_{2\,(g)} \rightarrow C_6H_{12}O_{6\,(s)} + 6O_{2\,(g)}$$

AN ENDOTHERMIC REACTION: Photosynthesis in plants turns water and carbon dioxide into glucose, storing the sun's energy in the sugar's molecular bonds. Oxygen is given off as a byproduct. The Calvin cycle is a multistep process that uses the coenzymes (biological catalysts, see page 38) ATP and NADPH to transfer energy, converting them into ADP and NADP+ and back again.

15

Hydrogen & Helium

primal elements

HYDROGEN accounts for around 74% of all known matter in the universe and is a large part of most stars. The first and simplest atom, it consists of one proton orbited by one electron (*opposite top*).

Hydrogen gas is **DIATOMIC**, which means it is happiest when two atoms covalently bond to form one molecule, H_2. Highly explosive in air, it burns rapidly with oxygen to create water (*see page 15*). Under immense pressures and temperatures (e.g. in the cores of giant planets like Jupiter and Saturn) hydrogen becomes metallic.

HELIUM, the second element in the periodic table, has two protons and two neutrons, making it four times heavier than hydrogen. With two electrons that completely fill the $1s$-orbital, it is the first *noble gas* and rarely reacts with other elements (*see page 34*). Despite being around 24% of all matter, it was unknown on Earth until 1870 when it was identified through the **SPECTROGRAPHIC ANALYSIS** of sunlight (*lower opposite*).

Beyond its common form, hydrogen has two isotopes, **DEUTERIUM**, with one neutron, and **TRITIUM**, with two. Tritium is rare and unstable, soon decaying into helium's light isotope **HELIUM-3** by changing a neutron into a proton via radioactive beta decay (*see page 42*).

Although the first members of the **s-BLOCK**, hydrogen and helium are found on opposite sides of the table, and are often considered separately due to their distinctive $1s$-orbital chemistry (*below*).

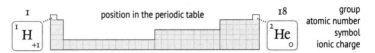

16

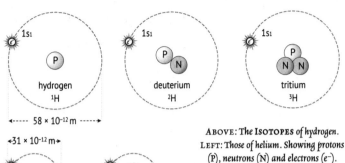

hydrogen
¹H

deuterium
²H

tritium
³H

←---- 58 × 10⁻¹² m ----→

←31 × 10⁻¹² m→

helium, ⁴He

helium–3, ³He

ABOVE: *The* ISOTOPES *of hydrogen.*
LEFT: *Those of helium. Showing protons*
(P), *neutrons* (N) *and electrons* (e⁻).
Helium's double proton charge pulls its
electrons nearer to the nucleus, making
it significantly smaller than hydrogen.
Note internal scales are exaggerated.

RIGHT: **A HYDROGEN FUEL CELL**:
Hydrogen is split into ions and electrons
at the platinum anode. A potassium
hydroxide electrolyte allows H⁺ ions
(protons) to pass, but blocks electrons
which travel round the circuit to power
the load. At the nickel cathode, the H⁺
ions and electrons are recombined along
with oxygen to make water:

$$4H^+ + 4e^- + 2O_2 \rightarrow 2H_2O$$

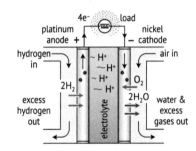

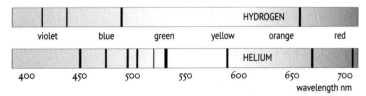

ABOVE: **EMISSION SPECTRA**: *Each element emits and absorbs at light unique frequencies, like*
a fingerprint. Analysing the light from compounds or distant stars can reveal their composition.

ALKALI & ALKALINE-EARTH METALS
the violent world of the s-block

The periodic table's first column is home to Group 1, the **ALKALI METALS**. Soft and silvery-white, they are highly **ELECTROPOSITIVE**, enthusiastically losing their one outer s-orbital electron to form singly charged +1 ions.

Working down the group, lithium (Li) is the lightest metal and floats and fizzes in water as it reacts. Sodium (Na) is even more reactive, it can boil water as it oxides, and is often found bonded with chlorine as common salt (NaCl). Potassium (K) oxidizes rapidly in air and bursts into flames when wet (*opposite centre*). Cæsium (Cs), the most electropositive element, and rubidium (Rb) both explode on contact with air. Lastly, francium (F) is both violently reactive and highly radioactive.

The neighbouring Group 2 **RARE EARTH METALS** are beryllium (Be), magnesium (Mg), calcium (Ca), strontium (Sr), barium (Ba), and radium (Ra). Only marginally less electropositive than Group 1, they readily lose their pair of outer electrons to form +2 ions. They are also denser with higher melting and boiling points.

A wire dipped in compounds of these elements will produce characteristic colours when held in a flame, as excited electrons jump between orbitals, losing a specific amount of energy as **PHOTONS** (packets of light)of a particular wavelength (hence the colour) on the way back to their usual state (*lower opposite*).

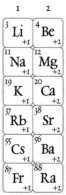

s-block

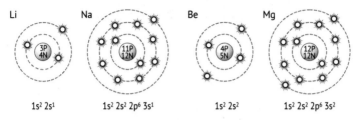

Li	Na	Be	Mg
3P 4N	11P 12N	4P 5N	12P 12N
$1s^2\ 2s^1$	$1s^2\ 2s^2\ 2p^6\ 3s^1$	$1s^2\ 2s^2$	$1s^2\ 2s^2\ 2p^6\ 3s^2$

ABOVE: ALKALI METALS, like lithium and sodium (left), and RARE EARTH METALS, such as beryllium and magnesium (right), readily lend their outer electrons to leave full orbital shells.

LEFT: POTASSIUM oxidizies furiously when dropped in water, providing enough heat to ignite the hydrogen gas given off, to form a basic, alkaline solution of potassium hydroxide.

$$2K_{(s)} + H_2O_{(l)} \rightarrow 2KOH_{(aq)} + H_{2(g)}$$

GROUP I

LITHIUM - carmine Red
SODIUM - yellow
POTASSIUM - lilac
RUBIDIUM - no colour
CAESIUM - no colour

GROUP II

BERYLLIUM - no colour
MAGNESIUM - no colour
CALCIUM - brick Red
STRONTIUM - crimson Red
BARIUM - apple Green

1. Place a sample in a flame

2. Energy from the flame excites an electron, which jumps to a higher orbital.

3. The electron falls back, emitting the excess energy as a photon of light.

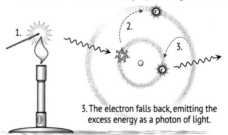

ABOVE: FLAME TESTING Groups I & II: Electrons need an exact amount of energy to jump between orbitals, which varies between atoms. The energy of a photon is directly related to the wavelength of light it emits, thus each energy jump creates photons of a particular colour.

THE P-BLOCK
metals, metalloids, & non-metals

Boron, the fifth element, introduces the P-BLOCK, as electron pairs fill three new double-teardrop *p*-orbitals at right angles to each other (*see page 11*). At room temperature and pressure, *p*-block elements can be solid (carbon and aluminium), liquid (bromine), or gas (nitrogen and chlorine), depending on the balance of interatomic and intermolecular forces.

The bottom left of the *p*-block mostly shines with metals. Ductile, malleable, and conductive because of their footloose outer electrons, metals can be stretched into wires, squashed into sheets, or combined into alloys. Left to right across the block, metals give way to non-metals. These tend to be gases, liquids, or dull brittle solids that are poor conductors of heat and electricity. Many of the elements essential to life are found here, such as carbon, oxygen, and nitrogen (*opposite top*), whose numerous compounds form the backbone of organic chemistry.

Between lie the METALLOIDS, ambiguous elements with aspects of both metals and non-metals (*shaded, right*). The SEMICONDUCTORS, boron, silicon, germanium, and arsenic, are useful in electronic components due to their

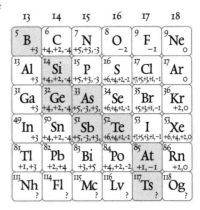

p-block

20

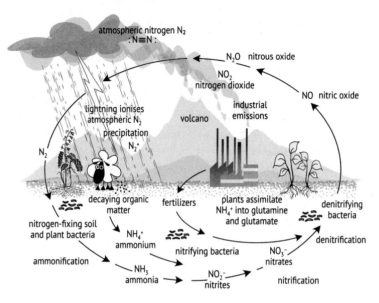

atmospheric nitrogen N_2
$:N\equiv N:$

N_2O nitrous oxide

NO_2
nitrogen dioxide

NO nitric oxide

lightning ionises
atmospheric N_2

volcano

industrial
emissions

precipitation
N_2^+

N_2

decaying organic
matter

fertilizers

plants assimilate
NH_4^+ into glutamine
and glutamate

denitrifying
bacteria

nitrogen-fixing soil
and plant bacteria

NH_4^+
ammonium

nitrifying bacteria

NO_3^-
nitrates

denitrification

ammonification

NH_3
ammonia

NO_2^-
nitrites

nitrification

ABOVE: THE NITROGEN CYCLE: Critical to life, nitrogen is found in all amino acids and proteins. To become bio-available, nitrogen-fixing bacteria must first break the strong N_2 triple bond. The nitrogen then passes through a series of biogeochemical processes as it recycles.

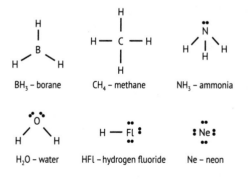

BH_3 – borane

CH_4 – methane

NH_3 – ammonia

H_2O – water

HFl – hydrogen fluoride

Ne – neon

LEFT: LEWIS STRUCTURES representing simple p-block molecules as the 2p-orbitals fill. Atoms beyond boron are surrounded by an octet of outer valence electrons. Lines represent bonded pairs of electrons. Unbonded lone pair electrons are shown as double dots.

CARBON & SILICON
organic electronic rock

Twenty-three percent of your body is CARBON, the sixth element. It forms the core of the many proteins within living cells, and is the basis of the oils and plastics made from fossil fuels. A non-metal, neither electropositive or negative, carbon easily combines with many elements. It can form long chains or rings, with double and triple held by multiple that smear bonds electrons between atoms (*see next page*).

Carbon atoms may be arranged in several forms, or ALLOTROPES. In diamond crystals, every atom bonds to four others in an incredibly hard tetrahedral grid (*opposite top left*); whereas in graphite, the soft crystalline material found in pencils, flat planes of carbon rings slide easily over each other, each atom joining to three others, the π-bonds enabling it to conduct electricity (*opposite top right*). Other allotropes include spherical FULLERENES, hollow NANOTUBES, and planar GRAPHENE, all of which have intriguing structural and conductive properties.

Below carbon in Group 14 is SILICON, the eighth most common element, often found combined with oxygen as SILICA (silicon dioxide, SiO_2), or with other oxides as SILICATES. Silicon compounds cover the Earth as rock, sand, and fine-grained clays—the crust is 60% aluminium silicate FELDSPARS ($KAlSi_3O_8$-$NaAlSi_3O_8$-$CaAl_2Si_2O_8$), and 20% quartz (crystallized SiO_2), whilst OLIVINE $(Mg,Fe)_2SiO_4$ makes 50% of the mantle. A metalloid semi-conductor, silicon can selectively transfer electrons, when doped with gallium or arsenic to alter its conductive properties, making silicon crystals an ideal substrate for electronic components like diodes, transistors and microchips (*lower, opposite*).

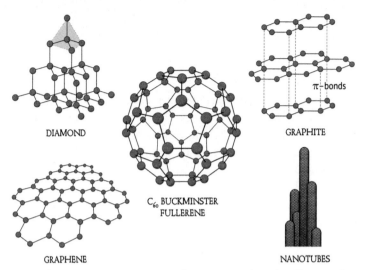

DIAMOND

GRAPHITE

π-bonds

C$_{60}$ BUCKMINSTER
FULLERENE

GRAPHENE

NANOTUBES

ALLOTROPES OF CARBON: From the hardness of diamond to the softness of graphite, pure carbon can take many forms. Natural fullerines, consisting of hundreds of atoms, can be found in soot.

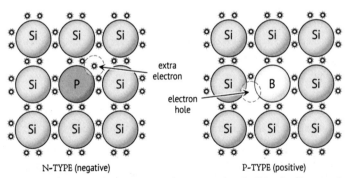

extra
electron

electron
hole

N-TYPE (negative)

P-TYPE (positive)

SILICON DOPING: In electronic transistors, other atoms are introduced into the silicon crystal lattice to change its properties. In p-type semiconductors, phosphorus' extra electron becomes a mobile charge carrier. Conversely, in n-type, boron creates a hole, accepting an extra electron.

ORGANIC CHEMISTRY
chains & rings

Carbon's versatile bonding abilities let it make millions of compounds with a huge variety of structures. ORGANIC CHEMISTRY focuses on carbon's interactions with hydrogen, often in combination with other elements such oxygen, nitrogen, sulfur, phosphorus and halogens.

HYDROCARBONS consist of only hydrogen and carbon, and form the basis of fossil fuels (*opposite upper left*). Carbon's ability to share bonds and create lattices allows it to POLYMERIZE, joining simple MONOMER molecules together into long chains of repeating units. POLYMERS create the fantastic plastics that are ubiquitous in today's world (*opposite upper right*).

UNSATURATED molecules have a double or triple carbon bond, which readily shifts allegiance to accommodate additional atoms or molecules. The multiple bond can occur in different places along a molecule's carbon backbone leading to related ISOMERS with the same formula but different behaviours. Molecules become SATURATED when all the multiple bond sites are taken and reduced to single bonds (*see below*).

AROMATIC compounds are based on a ring of carbon atoms linked by multiple overlapping bonds (*opposite lower*). Cyclic molecules are used industrially as the building blocks of synthetic drugs, polymers and pesticides. Aromatic molecules are also essential to biochemistry, where they form the heart of many amino acids and proteins.

C_4H_8 but-1-ene C_4H_8 but-2-ene C_4H_{10} butane

.... isomers ....

.... UNSATURATED ....   SATURATED

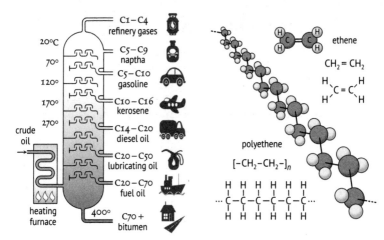

FRACTIONAL DISTILLATION: *Crude oil, a mix of hydrocarbons, is separated into different grades. Light gases rise the highest, whilst heavy bitumens sink to the bottom.*

POLYMERIZATION: *Ethene's double bond breaks to form the saturated singly bonded backbone of polyethene. One polymer molecule may contain 20,000 monomers.*

In the fractional distillation diagram:

20°C — C1–C4 refinery gases
70° — C5–C9 naptha
120° — C5–C10 gasoline
170° — C10–C16 kerosene
270° — C14–C20 diesel oil
— C20–C50 lubricating oil
— C20–C70 fuel oil
400° — C70 + bitumen

crude oil

heating furnace

ethene

$CH_2 = CH_2$

polyethene

$[-CH_2-CH_2-]_n$

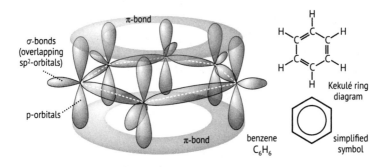

σ-bonds (overlapping sp²-orbitals)

π-bond

p-orbitals

π-bond

benzene C_6H_6

Kekulé ring diagram

simplified symbol

The **BENZENE RING** is a planar structure of multiply bonded carbon atoms. The σ-bonds between the carbon and hydrogen atoms, combined with π-bonds from the shared carbon p-orbitals, form an overlapping delocalized bond system that makes cyclic molecules very stable.

FUNCTIONAL GROUPS
names & numbers

Organic molecules are named using a standardized system. The number and arrangement of atoms in the carbon skeleton is shown by a prefix; *meth–* for one carbon atom, *eth–* for two, *prop–* for three etc. Compounds are classified by their FUNCTIONAL GROUP, a molecular subunit which reacts in a particular way, indicated by a suffix: *-ane, -ene, -yne* etc. When several hydrocarbons branch off a chain, the branches are numbered according to their location on the longest molecule. An "R" signifies a hydrogen atom or hydrocarbon group (*opposite*).

Fermenting fruit produces ALCOHOLS (with an –OH functional group, *shaded below*), mostly *ethanol*, with two carbon atoms (C_2H_5OH), which makes a pleasant drink. Alas, some brews also contain traces of single carbon *methanol* (CH_3OH), which is toxic. The alcohols are metabolized by the body into ALDEHYDES (–CHO) resulting in a hangover; however the pain caused by *ethanal* is tame compared to the odious effects of *methanal*.

The aldehydes are further processed into CARBOXYLIC ACIDS (–COOH), *ethanoic acid* (CH_3COOH) and *methanoic acid* (CHOOH). Again just one carbon atom makes all the difference—ethanoic (acetic) acid is the tasty tang of vinegar, whilst methanoic (formic) acid has a nasty sting and is squirted by ants to defend themselves.

methanol	ethanol	ethanal	ethanoic acid
CH_3OH	C_2H_5O	C_2H_4OH	C_2H_3COOH

Group Name:	Functional Group:	Suffix Name:	First Member Structure:	Name:	General Formula:
ALKANE	$R - CH_3$	-ane	$H-\overset{\overset{H}{\mid}}{C}-H$	methane	C_nH_{2n+2}
ALKENE	$\underset{R}{\overset{H}{\diagdown}}C=\underset{R'}{\overset{H}{\diagup}}$	-ene	$\underset{H}{\overset{H}{\diagdown}}C=\underset{H}{\overset{H}{\diagup}}$	ethene	C_nH_{2n}
ALKYNE	$R-C\equiv C-R'$	-yne	$H-C\equiv C-H$	ethyne	C_nH_{2n-2}
ALCOHOL	$R-\overset{\overset{OH}{\mid}}{\underset{\underset{H}{\mid}}{C}}-R'$	-anol	$H-\overset{\overset{OH}{\mid}}{\underset{\underset{H}{\mid}}{C}}-H$	methanol	$C_nH_{2n+1}OH$
ALDEHYDE	$R-\overset{\overset{O}{\mid\mid}}{C}-H$	-anal	$H-\overset{\overset{O}{\mid\mid}}{C}-H$	methanal	$C_nH_{2n-1}OH$
KETONE	$\underset{\underset{O}{\mid\mid}}{\overset{R}{\diagdown}}C{\diagup}^{CH_3}$	-anone	$\underset{\underset{O}{\mid\mid}}{\overset{H}{\diagdown}}C{\diagup}^{CH_3}$	methanone	$C_nH_{2n}O$
CARBOXYLIC ACID	$R-\overset{\overset{O}{\mid\mid}}{C}-OH$	-anoic acid	$H-\overset{\overset{O}{\mid\mid}}{C}-OH$	methanoic acid	$R-COOH$
AMINE	$R-NH_2$	-anamine	$CH_3-N{\overset{\diagup H}{\diagdown}}_H$	methylamine	$R_{3-n}NH_n$
ESTER	$R{\diagdown}\overset{\overset{O}{\mid\mid}}{C}{\diagup}_{OR'}$	-ate	$H-C{\overset{\diagup O}{\diagdown}}_{OCH_3}$	methyl methanoate carboxylic acid + alcohol	
AMIDE	$R{\diagdown}\overset{\overset{O}{\mid\mid}}{C}{\diagup}_{NHR'}$	-ide	$H{\diagdown}\overset{\overset{O}{\mid\mid}}{C}{\diagup}_{NH_2}$	methanamide carboxylic acid + amine	
ETHER	$R-O-R'$	-oxy...-ane	CH_3-O-CH_3	methoxy methane	

CYCLOPROPANE

$H-\overset{\overset{\overset{H}{\mid}}{C}}{\diagup\diagdown}\underset{H}{\overset{H}{\mid}}C{-}\underset{H}{\overset{H}{\mid}}C-H$

BENZENE RING

$H-\overset{\overset{\overset{H}{\mid}}{C}}{\underset{\underset{}{}}{}} ... C-H$ or ⬡ or ⬡

BRANCHED ALKANES: branch name changes to -yl

(CH₂—CH₃) ethyl branch

$\underset{1}{CH_3}\ \underset{2}{CH_2}\ \underset{3}{CH}\ \underset{4}{CH_2}\ \underset{5}{CH_2}\ \underset{6}{CH_3}$

3-ethyl hexane

(CH₃) methyl branch

$\underset{1}{CH_3}\ \underset{2}{CH}\ \underset{3}{CH_2}\ \underset{4}{CH_2}\ \underset{5}{CH_3}$

2-methyl pentane

ABOVE: **CARBON COMPOUNDS:** Prefix (no. of carbon atoms): meth-1; eth-2; prop-3; but-4; pent-5; hex-6; hept-7; oct-8; non-9; dec-10. The symbols R and R′ denote either a hydrogen atom, or a hydrocarbon side chain.

OXYGEN & SULFUR
over and underworlds

OXYGEN is the universe's third most abundant element. Intensely reactive, it exerts a strong pull on other atoms to gain the two electrons needed to fill its outer orbitals. Oxygen reactions play a central role in chemistry and are fundamental to life, being involved in everything from sending nerve impulses to metabolizing food and energizing cells.

On Earth, free oxygen is found in the atmosphere as a diatomic gas (O_2), and higher up as triatomic OZONE (O_3), which absorbs ultraviolet light. Earth's crust is mostly OXIDES; 60% silicon dioxide (SiO_2, as sand, quartz or other silicates), 35% magnesium oxide (MgO), and nearly 9% iron(II) oxide (FeO). Another oxide, water (H_2O) covers 71% of the surface, and is involved in many geochemical cycles (*opposite top*).

SULFUR lies beneath oxygen in the periodic table, and somewhat mirrors its chemistry. Also known as brimstone, it forms a pale yellow brittle solid with many allotropes (*lower, opposite*).

Sulfur is less electronegative than oxygen, so rotten-smelling hydrogen sulfide (H_2S) exhibits little hydrogen bonding and has different properties to water. Burning produces sulfur dioxide (SO_2), which combines with water in clouds to fall as sulfuric acid rain (H_2SO_4).

Sulfur's ability to build bridges between molecules is useful to living cells and it is found in several amino acids, acting as a structural glue holding proteins together (*right*).

cystine

disulfide bridge

cystine

28

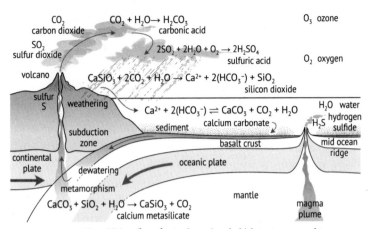

CO_2
carbon dioxide

$CO_2 + H_2O \rightarrow H_2CO_3$
carbonic acid

O_3 ozone

SO_2
sulfur dioxide

$2SO_2 + 2H_2O + O_2 \rightarrow 2H_2SO_4$
sulfuric acid

O_2 oxygen

volcano

$CaSiO_3 + 2CO_2 + H_2O \rightarrow Ca^{2+} + 2(HCO_3^-) + SiO_2$
silicon dioxide

sulfur
S

weathering

$Ca^{2+} + 2(HCO_3^-) \rightleftharpoons CaCO_3 + CO_2 + H_2O$

H_2O water

sediment

calcium carbonate

H_2S hydrogen sulfide

subduction zone

basalt crust

mid ocean ridge

continental plate

dewatering

oceanic plate

metamorphism

$CaCO_3 + SiO_2 + H_2O \rightarrow CaSiO_3 + CO_2$
calcium metasilicate

mantle

magma plume

EARTH CYCLES: *The oxidising effects of water drawn into the high temperature and pressure environment of subduction zones creates new generations of metamorphic rock. Meanwhile, life thrives around mid-ocean ridges on energy extracted by bacteria from volcanic hydrogen sulfide.*

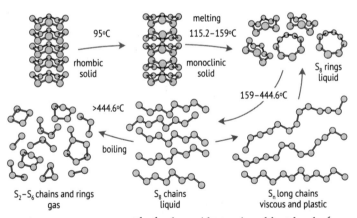

$95^{\circ}C$

rhombic solid

melting
$115.2-159^{\circ}C$

monoclinic solid

S_8 rings liquid

$159-444.6^{\circ}C$

$>444.6^{\circ}C$

boiling

S_2-S_6 chains and rings gas

S_8 chains liquid

S_n long chains viscous and plastic

SOME ALLOTROPES OF SULPHUR: *Often found as an eight atom ring, sulphur takes other forms, depending on temperature and pressure. At extremely high pressures, it even becomes a metal.*

WATER, ACIDS & BASES
concerning dihydrogen monoxide

WATER (H_2O) is abundant across the universe. A seemingly simple molecule, it has a treasure trove of unusual properties. In particular, water is POLARIZED—the oxygen's lone pair electrons pull at the hydrogen atoms to give a slight charge across the molecule. This causes loose HYDROGEN-BONDS, which create the shifting lattices of liquid water, its strong surface tension, and the structure of ice (*opposite top*).

Water's polarization makes it a particularly effective solvent, pulling apart molecules in solution to form charged ions. It even dissolves itself, with some water molecules losing protons (hydrogen ions, H^+) to others, creating a solution of hydronium (H_3O^+) and hydroxide (OH^-) ions.

ACIDS are compounds that are proton donors in aqueous solution and will corrode metals to liberate hydrogen. Conversely, BASES are soapy, bitter compounds that act as proton acceptors. SALTS are formed when an acid and a base combine to neutralize each other (*lower opposite*).

The caustic Group I and II alkalis are particularly BASIC, along with some metal oxides, hydroxides, and amines. Not all acids need water— LEWIS acids and bases accept/donate electron pairs in other solvents.

The strength of an acid or base is determined by how completely it dissociates into ions, and is measured using the *p*H SCALE, the negative logarithm of a solution's hydrogen ion concentration (*below*).

*p*H	strongly acidic							neutral						strongly basic	
	O	I	2	3	4	5	6	7	8	9	IO	II	I2	I3	I4
	red			orange				yellow			green			blue	

colour of litmus test paper

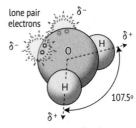

The **Water Molecule** (H_2O): the electrical charge asymmetry (δ^+, δ^-) creates a polarized molecule that can form loose hydrogen bonds.

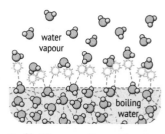

Breaking **Hydrogen Bonds** takes energy, resulting in water's relatively high boiling point and heat of vaporization (40.65 kJmol^{-1}).

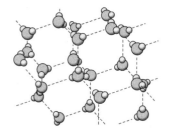

Hexagonal Crystals of ice I_h form at $0°C$ at atmospheric pressure. Ice has at least 18 phases depending on temperature and pressure, even becoming metallic under intense pressures.

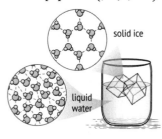

Unusually **Liquid Water** is denser than solid, allowing ice to float. Hydrogen bonds form a fluctuating structure which expands and becomes more ordered as water freezes.

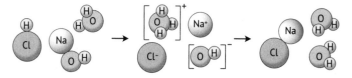

$$HCl_{(g)} + NaOH_{(s)} + H_2O_{(l)} \rightarrow H_3O^+_{(aq)} + Cl^-_{(aq)} + Na^+_{(aq)} + OH^-_{(aq)} \rightarrow NaCl_{(aq)} + 2H_2O_{(l)}$$

A Neutralization Reaction: hydrochloric acid (HCl) and sodium hydroxide (NaOH), a strong acid and base, dissociate completely into ions when in solution, and combine in a double displacement reaction to form common salt (sodium chloride, NaCl) and water.

REDOX REACTIONS
an ox loses what the red cat gains

Rusty cars, log fires and browning apples are examples of REDOX (*reduction-oxidation*) reactions, where there is a change in OXIDATION NUMBER—the number of electrons a covalently bonded electronegative atom attracts from a more electropositive one (*opposite top left*).

Redox reactions unfold simultaneously as complementary halves: In the OXIDATION part a REDUCTANT (reducing agent) loses electrons and becomes OXIDIZED, leading to an increase in its oxidation number. In the REDUCTION part an OXIDANT (oxidizing agent) gains electrons and is REDUCED (*below*), leading to an decrease in its oxidation number.

There are two main types: ATOM TRANSFERS, e.g. when hydrogen atoms are shuffled to form saturated organic molecules, or when oxygen combines with another element to form an oxide (*lower, opposite*). The term 'oxidation' was originally used solely in this latter context, however it now applies to all reactions involving electron loss.

ELECTRON TRANSFERS direct the processes of electroplating (*opposite top right*) and electrolysis (*page 37*). An example is the 'sacrificial anode' used to prevent underwater corrosion—by bolting a piece of zinc or aluminium to a ship's hull, the oxygen in the water will oxidize the more electropositive metal, dissolving it in preference to the steel.

A handy mnemonic for remembering these reactions is OIL RIG: *Oxidation Is (electron) Loss, Reduction Is Gain.* Or alternatively, AN OX LOSES WHAT THE RED CAT GAINS: *ANode-OXidation-LOSES electrons. REDuction-CAThode-GAINS electrons.*

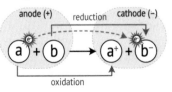

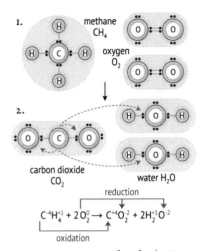

At the anode: $Ag_{(s)} - e^- \longrightarrow Ag^+_{(aq)}$
reduction

At the cathode: $Ag^+_{(aq)} + e^- \longrightarrow Ag_{(s)}$
oxidation

reduction

$$C^{-4}H_4^{+1} + 2O_2^0 \longrightarrow C^{+4}O_2^{-2} + 2H_2^{+1}O^{-2}$$

oxidation

Burning Methane: 1. *The carbon in CH$_4$ attracts shared electrons. 2. On combustion, the oxygen attracts electrons from both carbon and hydrogen (oxidation numbers shown in grey).*

Electroplating: *A battery drives electrons from anode to cathode; silver ions dissolve into solution at the anode and are deposited onto the metal spoon cathode, coating it with silver.*

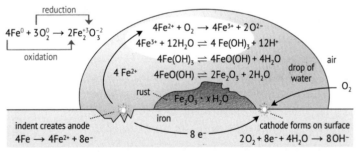

reduction

$$4Fe^0 + 3O_2^0 \longrightarrow 2Fe_2^{+3}O_3^{-2}$$

oxidation

$4Fe^{2+} + O_2 \longrightarrow 4Fe^{3+} + 2O^{2-}$

$4Fe^{3+} + 12H_2O \rightleftharpoons 4Fe(OH)_3 + 12H^+$

$4Fe(OH)_3 \rightleftharpoons 4FeO(OH) + 4H_2O$

$4FeO(OH) \rightleftharpoons 2Fe_2O_3 + 2H_2O$

$4Fe^{2+}$

rust $Fe_2O_3 \cdot xH_2O$

drop of water

air

O_2

indent creates anode
$4Fe \longrightarrow 4Fe^{2+} + 8e^-$

iron

$8e^-$

cathode forms on surface
$2O_2 + 8e^- + 4H_2O \longrightarrow 8OH^-$

Rusting *begins when iron is oxidised into Fe^{2+} ions. The freed electrons reduce oxygen to form hydroxide (OH$^-$) ions. A further redox reaction transforms the Fe^{2+} ions into Fe^{3+} ions, which pass through a series of steps to finally precipitate as hydrous iron(III) oxide — rust.*
Note: *Oxidation numbers are shown as Fe(III) or as Fe^{+3}; Ionic charges are written Fe^{3+}.*

HALOGENS & THE NOBLE GASES
ups and downs at period's end

The most vigorous and, by contrast, the most inert elements are found next door to each other in the final two columns of the periodic table.

Highly **ELECTRONEGATIVE**, the Group 17 **HALOGENS** are one electron short of a full outer shell and aggressively form compounds by grabbing an electron to fill it (*lower, opposite*). All elements, except helium, neon, and argon, react with a halogen to form **HALIDES** (*opposite centre*).

The ninth element, and easily the most electronegative, is **FLUORINE** (*opposite top left*), a pale green-yellow diatomic gas which combines fanatically with almost anything, attacking compounds to form **FLUORIDES**. Below this, **CHLORINE** ranks as third most electronegative after oxygen, making it ideal for chemical reagents, bleaches, disinfectants and poisons. Paradoxically, it is also essential to life—the chlorine in common salt (NaCl) contributes to cellular function and powers digestion as hydrochloric acid (HCl).

With one more proton and one more electron, the period ends with solipsistic Group 18. Their full orbitals mean they are generally unreactive. However, xenon will, with effort, combine with feisty fluorine (XeF_6) and its neighbour, oxygen (XeO_4). A few helium and krypton compounds also exist, so in 1957 the old name for the group, the **INERT GASES**, was changed to the slightly less lazy **NOBLE GASES** (*opposite top right*).

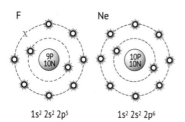

F   Ne

1s² 2s² 2p⁵   1s² 2s² 2p⁶

$1s^2\ 2s^2\ 2p^5$ $1s^2\ 2s^2\ 2p^6$

HALOGENS like fluorine (left) are highly
reactive due to being one electron short (x)
of a full outer shell, unlike the unreactive
NOBLE GASES, such as neon (right).

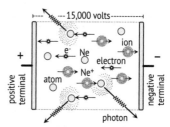

A NEON SIGN: Applying high voltage across
a gas-filled tube strips electrons from atoms.
Speeding ions and electrons collide with atoms
shedding excess energy as photons of light.

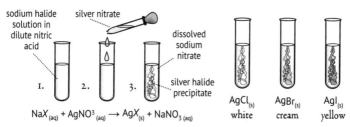

$$NaX_{(aq)} + AgNO_{3\ (aq)} \rightarrow AgX_{(s)} + NaNO_{3\ (aq)}$$

UNKNOWN HALIDE IONS in solution can be identified by precipitating a silver halide: 1. Acidify
the solution with dilute nitric acid. 2. Add silver nitrate. 3. The solid precipitate's colour reveals the
halogen—silver chloride is milk white, silver bromide is cream and silver iodide is butter yellow.

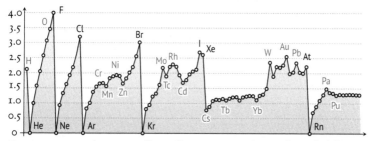

ELECTRONEGATIVITY is a measure of how easily an atom attracts electrons: the reactive halogens
occupy the peaks whilst the noble gases (with the exception of xenon) sit in the valleys.

THE TRANSITION METALS

yes my precious

The next section of the periodic table, the **D-BLOCK**, begins at **SCANDIUM** (Sc) where the first of ten electrons begins to fill a fresh series of *3d*-orbitals *inside* the *4s* (*see page 7*). Known as the **TRANSITION METALS**, these elements can lose one or more electrons to form an array of brightly coloured compounds. Many are hard and very strong: tungsten (W) is the strongest metal, with chromium (Cr) a close second, and titanium (Ti) a lightweight yet durable choice for aerospace.

Transition metals share a similar crystal structure, allowing them to be blended into **ALLOYS**. Copper (Cu) melds with tin (Sn) to form **BRONZE**, and with zinc (Zn) to make **BRASS**. Equal amounts of chromium, cobalt, and nickel make exceptionally tough CrCoNi. Mercury (Hg), the only metal liquid at room temperature, forms malleable alloys called **AMALGAMS**.

Adding a little carbon to iron turns it into steel (*opposite top left*), which is fortified with a pinch of vanadium (V), molybdenum (Mo), or chromium. Unpaired electrons in the transition metals' outer *d*-orbitals produce iron's **FERROMAGNETIC** properties, which are partly shared with nickel and cobalt. Most members of the d-block are **PARAMAGNETIC** and will be attracted to a magnetic field (*shaded, lower opposite*).

Copper (Cu), silver (Ag), and gold (Au) are known as 'coinage metals', their lustre and resistance to corrosion making them an ideal choice for crowns, coins, rings and other baubles. They are excellent conductors of electricity and heat, with many applications in electronics and optics (*opposite top right*).

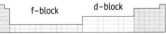

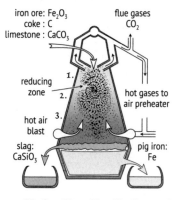

iron ore: Fe_2O_3
coke : C
limestone : $CaCO_3$

flue gases
CO_2

reducing zone

1.

hot gases to air preheater

2.

hot air blast

3.

slag: $CaSiO_3$

pig iron: Fe

1: $3Fe_2O_3 + CO \rightarrow CO_2 + 2Fe_3O_4$ 450°C

2: $Fe_3O_4 + CO \rightarrow CO_2 + 3FeO$ 580°C

3: $\begin{cases} FeO + CO \rightarrow CO_2 + Fe \\ FeO + C \rightarrow CO + Fe \end{cases}$ 700°C

SMELTING IRON in a blast furnace: Iron ore, iron(III) oxide, is reduced to iron, oxidising carbon to form carbon dioxide. The reaction sequence progresses as temperature increases.

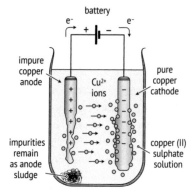

battery

e^- e^-

+   −

impure copper anode

Cu^{2+} ions

pure copper cathode

impurities remain as anode sludge

copper (II) sulphate solution

At the anode: $Cu(s) \rightarrow Cu^{2+}(aq) + 2e^-$
oxidation

At the cathode: $Cu^{2+}(aq) + 2e^- \rightarrow Cu(s)$
reduction

PURIFYING COPPER by electrolysis: electrons flow from the anode, dissolving copper ions into solution. Attracted to the cathode, the ions regain their electrons, depositing pure copper.

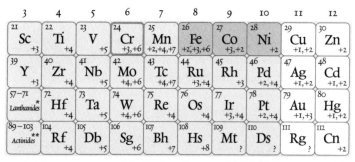

3	4	5	6	7	8	9	10	11	12
21 Sc +3	22 Ti +4	23 V +5	24 Cr +3,+6	25 Mn +2,+4,+7	26 Fe +2,+3,+6	27 Co +3,+2	28 Ni +2	29 Cu +1,+2	30 Zn +2
39 Y +3	40 Zr +4	41 Nb +5	42 Mo +4,+6	43 Tc +4,+7	44 Ru +3,+4	45 Rh +3	46 Pd +2,+4	47 Ag +1,+2	48 Cd +1,+2
57–71 Lanthanides*	72 Hf +4	73 Ta +5	74 W +4,+6	75 Re +4	76 Os +4	77 Ir +3,+4	78 Pt +2,+4	79 Au +1,+3	80 Hg +1,+2
89–103 Actinides**	104 Rf +4	105 Db +5	106 Sg +6	107 Bh +7	108 Hs +8	109 Mt ?	110 Ds ?	111 Rg ?	112 Cn +2

THE TRANSITION METALS: Dark shading shows ferromagnetism; light shading paramagnetism. Chromium is antiferromagnetic at room temp. Many transition compounds are also paramagnetic.
*Elements 57-71, the lanthanides, and **89-103, the actinides, form the f-block (see page 40).

CATALYSTS
the need for speed

CATALYSTS are substances that speed up reactions, yet remain unchanged by them. A reaction proceeds by the breaking and rebuilding of bonds which needs a certain level of ACTIVATION ENERGY. Catalysts decrease this threshold by providing an easier route, enabling atoms to rearrange at lower temperatures and/or pressures (*opposite top*). They can also be selective, increasing yields whilst reducing unwanted products.

Several transition metal ions exist in a number of oxidation states, making them effective industrial catalysts—the HABER PROCESS uses an iron based catalyst to fix atmospheric nitrogen with hydrogen to produce ammonia. Long chain hydrocarbons are cracked into smaller, more useful subunits by ZEOLITES, porous minerals of aluminium, silicon and oxygen. Exhaust gases are cleaned with catalytic converters of rhodium, platinum, and palladium (*lower, opposite*).

Life is sustained by ENZYMES, complex folded chains of proteins often made of thousands of atoms (*right*). Evolving soon after life began on Earth, they act as biochemical catalysts increasing the rate of slow metabolic processes, sometimes by a factor of millions. All living things have them—some bacterial enzymes like PETASE can even digest plastic (*below*).

petase structure

| polyethylene terephthalate (PET) | mono(2-hydroxyethyl) terephthate | terephthalic acid | ethylene glycol |

POLYMER   MONOMERS

38

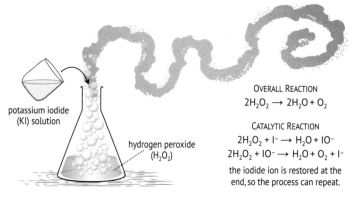

OVERALL REACTION

$$2H_2O_2 \rightarrow 2H_2O + O_2$$

CATALYTIC REACTION

$$2H_2O_2 + I^- \rightarrow H_2O + IO^-$$
$$2H_2O_2 + IO^- \rightarrow H_2O + O_2 + I^-$$

the iodide ion is restored at the end, so the process can repeat.

ABOVE: ELEPHANT TOOTHPASTE uses a potassium iodide catalyst to speed up the decomposition of hydrogen peroxide into water and oxygen, compressing a reaction that usually takes days into a few seconds. Adding a little soap traps the released oxygen, producing a rapidly growing foam tube.

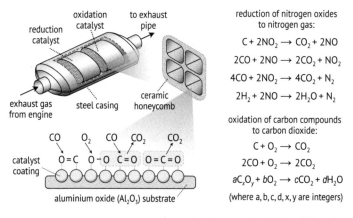

reduction of nitrogen oxides to nitrogen gas:

$$C + 2NO_2 \rightarrow CO_2 + 2NO$$
$$2CO + 2NO \rightarrow 2CO_2 + NO_2$$
$$4CO + 2NO_2 \rightarrow 4CO_2 + N_2$$
$$2H_2 + 2NO \rightarrow 2H_2O + N_2$$

oxidation of carbon compounds to carbon dioxide:

$$C + O_2 \rightarrow CO_2$$
$$2CO + O_2 \rightarrow 2CO_2$$
$$aC_xO_y + bO_2 \rightarrow cCO_2 + dH_2O$$

(where a, b, c, d, x, y are integers)

ABOVE: CATALYTIC CONVERTERS transform toxic exhaust gases into less harmful byproducts. Three-way converters employ two catalysts in series – rhodium / platinum to reduce nitrogen compounds and palladium / platinum to oxidise carbon, carbon monoxide and other hydrocarbons.

THE F-BLOCK & SUPERHEAVIES
rare earth metals and islands of stability

At LANTHANUM, a new *5d*-orbital begins to fill. Surprisingly, the next electron drops into a hidden *4f*-orbital taking the one from the *5d* with it. The *5d*-orbitals then have to wait for the *4f* to fill, apart from gadolinium, where an electron fleetingly jumps back to the *5d*. Although having room for fourteen electrons, the intricate *f*-orbitals are overshadowed by the outer shells and play little part in bonding (*opposite top and frontispiece*).

Spreading from cerium to lutetium, the LANTHANIDES form the first row of the F-BLOCK (*below, upper row*). Also known as the RARE-EARTH METALS, their subtly different chemical properties make them important in the manufacture of solar panels, batteries and catalysts.

Beneath the lanthanides, the ACTINIDES play a similar trick, with two electrons starting a *6d*-orbital only to quit and fill an inner *5f*-orbital instead (*below, lower row*). All the actinides are radioactive, with uranium being the last stable naturally formed element. Trace amounts of decaying neptunium and plutonium have also been found in uranium ores.

With the completion of the *5f* at lawrencium, a fourth transition series begins a *6d*-orbital. A number of short-lived, artificially synthesized, SUPERHEAVY elements with up to 118 protons have been fleetingly created in particle accelerators. Beyond, a hypothetical element 119 may herald the start of an as yet unrevealed *g*-orbital (*opposite lower*).

lanthanides	57 La +3	58 Ce +3,+4	59 Pr +3	60 Nd +3	61 Pm +3	62 Sm +3	63 Eu +2,+3	64 Gd +3	65 Tb +3	66 Dy +3	67 Ho +3	68 Er +3	69 Tm +3	70 Yb +3	71 Lu +3	
	89 Ac +3	90 Th +4	91 Pa +5	92 U +6	93 Np +5	94 Pu +4	95 Am +3	96 Cm +3	97 Bk +3	98 Cf +3	99 Es +3	100 Fm +3	101 Md +3	102 No +2	103 Lr +3	actinides

LANTHANIDES

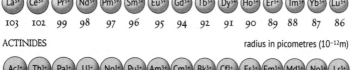

La³⁺	Ce³⁺	Pr³⁺	Nd³⁺	Pm³⁺	Sm³⁺	Eu³⁺	Gd³⁺	Tb³⁺	Dy³⁺	Ho³⁺	Er³⁺	Tm³⁺	Yb³⁺	Lu³⁺
103	102	99	98	97	96	95	94	92	91	90	89	88	87	86

ACTINIDES

radius in picometres (10⁻¹²m)

Ac³⁺	Th³⁺	Pa³⁺	U³⁺	Np³⁺	Pu³⁺	Am³⁺	Cm³⁺	Bk³⁺	Cf³⁺	Es³⁺	Fm³⁺	Md³⁺	No³⁺	Lr³⁺
112	109	108	107	106	105	104	103	102	101	100	99	99	98	98

LANTHANIDE AND ACTINIDE CONTRACTION: *The diffuse nature of the f-orbitals results in a low electron density, poorly shielding the outer electrons from the pull of the nucleus. This attraction increases with each additional proton causing a decrease in radii across the f-block.*

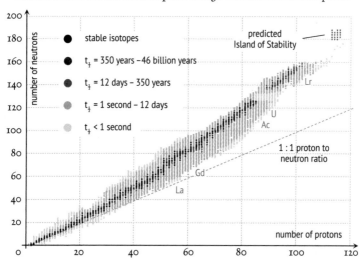

ABOVE: ISLANDS OF STABILITY: *Nuclei are stabilised by an interplay of forces. Protons repel each other, requiring neutrons to stabilise the atom; too many neutrons leads to neutron decay, changing a neutron into a proton, electron and antineutrino. An "island of stability", with about 114 protons and 184 neutrons, may harbour isotopes with significant lifetimes. Others might exist with 126 and 164 protons. Lifetimes are given in half-life ($t_{\frac{1}{2}}$), the time taken for half a sample's atoms to decay.*

RADIOACTIVITY
nuclear fizzicks

Held together by immensely strong forces, nuclei contain vast amounts of energy. Unstable nuclei rebalance through RADIOACTIVE emission, either by joining nucleons (*fusion*), or splitting them apart (*fission*). Radioactivity is measured in HALF-LIFE ($t_{1/2}$), the time it takes for half the atoms in a sample to decay—the less stable it is, the faster it disintegrates. Uranium-238 has a half-life of a 4.5 billion years, yet with ten neutrons less, uranium-228's half-life is only a fifth of a second (*opposite*).

Living things absorb the slightly radioactive carbon-14 that is generated by cosmic rays. At death, cells stop gathering the isotope, allowing archæologists to date organic matter by comparing the ratios of carbon-14, with a 5,730-year half-life, with that of stable carbon-12.

Beyond bismuth (Bi, *p.23*), all elements have RADIOISOTOPES that undergo α-(ALPHA) DECAY, the nucleus expelling a helium nucleus (an *α-PARTICLE, see page 7*). Thin clothing should shield you from α-particles. Too many neutrons in a nucleus causes β-(BETA) DECAY, where a neutron converts into a proton, shedding energy as an electron (a β-PARTICLE) and an antineutrino. Protective suits or 2 mm of aluminium stops these. Often produced alongside α- or β-decays are high-intensity photons, γ-(GAMMA) RAYS of ionizing electromagnetic radiation. Effective shielding requires several inches of lead or feet of concrete (*below*).

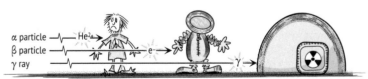

α particle —— He²⁺
β particle —— e⁻
γ ray

ABOVE: NUCLEAR FUSION *occurs when nuclei collide to form a heavier element. Here, hydrogen isotopes deuterium and tritium fuse into helium, releasing a neutron and 17.6 MeV of energy. For atoms up to iron, fusion releases energy. Beyond it requires energy, so elements heavier than iron can only be forged during extremely high energy supernovæ or in particle accelerators.*

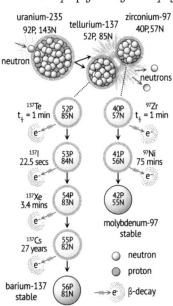

NUCLEAR FISSION: *Struck by a neutron, an atom of uranium-235 splits into two atoms (Te & Zr), expelling two neutrons. Each follows a decay chain until becoming a stable atom.*

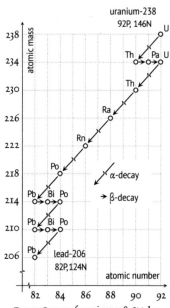

DECAY CHAIN *of uranium-238. It takes 4.5 billion years for half a sample of* ^{238}U *to transmute into stable lead-206 via a multistep series of alpha- and beta-decays.*

43

QUIRKY QUARKS
and curious quantum effects

Deep inside the NUCLEONS (protons and neutrons) energy and matter have such a close relationship that it's not easy to tell them apart.

Both nucleons are made of three QUARKS, indivisible sub-atomic particles which form the fundamental building blocks of the material universe, in this case UP and DOWN quarks. Up quarks have a $+\frac{2}{3}$ electric charge, whilst down quarks carry $-\frac{1}{3}$. Protons consist of two up quarks and one down quark, giving a total charge of $\frac{2}{3}+\frac{2}{3}-\frac{1}{3} = 1$. By contrast, a neutron has one up quark and two down quarks, whose charges cancel out, $\frac{2}{3}-\frac{1}{3}-\frac{1}{3} = 0$, leaving it neutral overall (*opposite top*).

Quarks are bound together in a nucleus by the STRONG FORCE, which is carried by fundamental particles known as GLUONS. The strong force has three charge polarities which can be loosely likened to the primary colours of light. Each quark in a nucleon needs to carry a different complementary COLOUR charge, which when combined results in an overall neutral colour charge across the particle (*opposite mid left*).

The strength of the strong force remains constant over its short range, regardless of distance. As a result, when quarks are pulled apart QUARK CONFINEMENT means it is more efficient to borrow energy from the quantum vacuum and spontaneously create quark/antiquark pairs than to break the gluon fields holding them together (*below*).

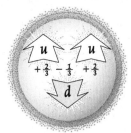

THE PROTON:
2 up quarks and 1 down, +1 charge.

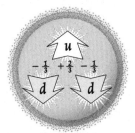

THE NEUTRON:
1 up quark and 2 down, 0 charge.

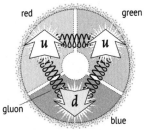

QUANTUM CHROMODYNAMICS:
Quarks bound by strong force GLUONS. The red, green and blue charges on individual quarks balance to give a neutral overall colour charge.

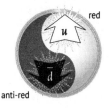

A POSITIVE PION (π^+).

MESONS, like the pion above, are $\frac{2}{3}$ the size of a nucleon, made from an up quark and an anti-down quark pair. Short-lived, they form in high energy collisions. Neutral overall colour charge is conserved through antiquarks having a complementary "anticolour" charge.

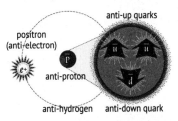

Most particles have an equal but oppositely charged ANTIMATTER twin (represented by a bar above the symbol).

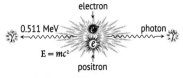

An electron and a positron collide. When matter meets anti-matter it ANNIHILATES, converting all of the mass into pure energy.

THE FOUR FORCES
holding it all together

Everything in the universe interacts with everything else through four universal forces, carried by four types of wave-particles called *gauge bosons*.

The WEAK force acts on subatomic particles over distances less than the diameter of a proton. It governs quark transformations and is resposible for radioactive decay. Carried by W and Z VECTOR BOSONS, the weak force also allows NEUTRINOS, very light fundamental particles, to perform their rare interactions with everyday matter (*see next page*).

The STRONG force binds quarks together by exchanging eight types of GLUON, the carriers of colour charge. Acting over a range covering the size of a nucleus, it only affects quarks and gluons.

The ELECTROMAGNETIC force creates the electric charge which attracts electrons to protons and mediates chemical reactions. Light, x-rays, radio waves, and microwaves are carried by PHOTONS which have dual electric and magnetic components that propagate perpendicularly (*below*). It is also the prime mover in matter/antimatter annihilations (*page 45*).

GRAVITY, although by far the weakest force, nevertheless operates over almost infinite distances, extending its grip over all matter. Finding a quantum explanation for gravity has proven difficult, though a carrier boson, the GRAVITON, has been tentatively described.

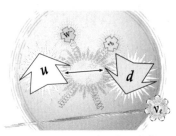

The WEAK FORCE acts on the scale of a nucleon. It is responsible for quark transformations and neutrino interactions.

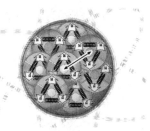

The STRONG FORCE operates over distances the size of a nucleus, holding the quarks, nucleons and nucleus together.

The ELECTROMAGNETIC FORCE works over all distances: it shapes electron orbits and the magnetic fields of planets and stars.

Although the weakest force, GRAVITY is felt across the universe, connecting every type of matter from galaxies to atoms.

FUNDAMENTAL FORCE:	RELATIVE STRENGTH:	RANGE: (m)
weak	10^{33}	10^{-18}
strong	10^{38}	10^{-15}
electromagnetic	10^{36}	∞
gravity	1	∞

A small magnet will easily pick up a nail, thereby overcoming the entire Earth's gravitational field, however gravity only attracts whilst electromagnetism can also repel – astronomical objects like planets and galaxies tend to be electrically neutral, making gravity the most influential force on cosmic scales.

FUNDAMENTAL FAMILIES
fermions, baryons & bosons

Ordinary matter is made of up and down quarks (which are **FERMIONS**) and electrons (which along with neutrinos are **LEPTONS**). At higher energies, as in cosmic rays, a heavier pair of quarks is found—**CHARM** and **STRANGE**, along with the **MUON** (a heavy electron) and **MUON-NEUTRINO**. A third even heavier family exists at extreme energies (*opposite top*). Each particle has an antimatter twin with the same mass, but opposite electric charge and **SPIN**, a quantum analogue of angular momentum.

Quarks, electrons and neutrinos have spin $\frac{1}{2}$ (*below*), and are thus **FERMIONS**, particles with half spins ($\frac{1}{2}, \frac{3}{2}, \frac{5}{2}, \ldots$). Fermions are bound by the **PAULI EXCLUSION PRINCIPLE**, which forbids duplication of quantum states, e.g. only electrons with opposite spin can share an atomic orbital. **BARYONS**, three quark composites like nucleons, are also fermions.

BOSONS are particles with integer spin ($0, 1, 2, \ldots$), examples being the force-carrying **GAUGE BOSONS**, and **MESONS** (quark/antiquark pairs, *see pages 45 & 58*). Bosons are not bound by the exclusion principle and can sit in the same quantum state, e.g. the coherent photons in a laser beam. Under certain conditions, even numbers of fermions can act together as bosons, giving rise to macroscopic quantum behaviours, such as zero-viscosity **SUPERFLUIDS**, or electrons that bind into **SUPERCONDUCTIVE** Cooper pairs which can flow with zero resistance.

FERMIONS ·················· BOSONS ··················

The SPIN $\frac{1}{2}$
Möbius Slug
twists twice
through 720°.

The SPIN 1
Tumble Snail
turns through
a full 360°.

The SPIN 2
Mirror Slug
has 180° half-
turn symmetry.

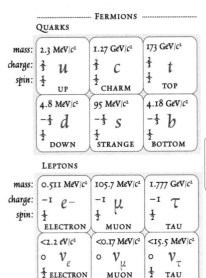

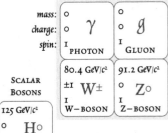

ABOVE: BOSONS *mediate the forces.* **CENTRE:** *The elusive* HIGGS BOSON *gives particles mass through interaction with an all pervasive Higgs field.* **LEFT:** FERMIONS. *Everyday matter particles are in column I. Column II particles occur in cosmic rays, whilst those in III are found at the highest energies.*

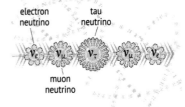

ABOVE: NEUTRINOS *only interact via the weak force and gravity, passing virtually unimpeded through ordinary matter. Oddly, they may exist as a quantum superposition of their three types, oscillating from one to another as they travel.*

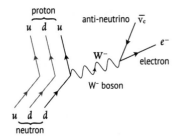

ABOVE: A FEYNMAN DIAGRAM, *representing the behaviour of subatomic particles. Here, a beta decay transforms a neutron into a proton, producing an electron and an anti-neutrino via a weak force interaction with a* W⁻ *boson.*

EXOTIC PARTICLES
subatomic siblings

Energetic cosmic rays from outer space—high-speed charged particles such as protons, helium nuclei, electrons or tiny amounts of antimatter—constantly hit the Earth, ionizing atoms in the upper atmosphere to create cascades of subatomic particles. Visible in the glowing polar auroras, branching streamers of HADRONS (quark composites of baryons and mesons) shed electrons, positrons, and γ-rays to cause further chain reactions, where short-lived exotic particles rapidly decay into more stable, lower energy ones (*opposite top*).

In experiments to recreate the extreme energies of the early universe, accelerators and colliders use magnetic fields to accelerate protons and other matter to near light speed, smashing them together to reveal their hidden internal structures. Sensitive detectors then trace the curling paths of hundreds of mostly short-lived rapidly transmuting particles (*below*). Some are higher energy versions, or *resonances*, of others, which can be linked into symmetrical patterns of families (*lower opposite*).

The quantum world still holds many surprises as nature continues to reveal her subtle harmonics.

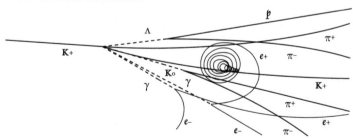

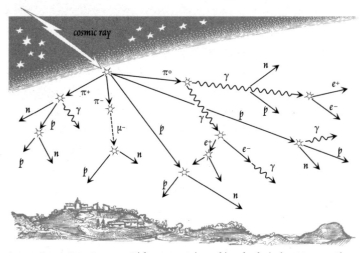

ABOVE: **COSMIC RAY CASCADE**. *High energy cosmic rays hit molecules in the upper atmosphere initiating secondary cascade chains of billions of subatomic particles, colliding and transmuting.*

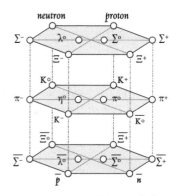

ABOVE: **EIGHTFOLD WAY OCTETS**: *hadron relationships charted as octets, correlating charges, spins, and other characteristics.*

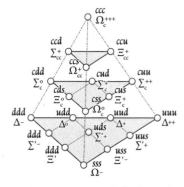

ABOVE: **SPIN 3/2 BARYONS**, *the family tree of three quark combinations. The triply charmed omega at the top has not yet been observed.*

QUANTUM THEORIES
interpretations of entanglements

Quantum theory is supremely successful at predicting the behaviour of matter on the smallest scales. Despite this, it is still not known *why* the application of matrix mechanics in the complex plane describes electron orbitals (*see pages 10-11*), nor *how* the states of an ENTANGLED pair of photons are instantly correlated regardless of distance (*below*), nor *why*, in the TWIN SLIT experiment, a single photon, atom, or molecule interferes as a *wave* (so passes through both slits) when unobserved, yet behaves as a *particle* (going through just one slit) when observed (*opposite top*).

Several 'interpretations' have been proposed. The COPENHAGEN INTERPRETATION suggests that observation *creates* microscopic reality— the act of measuring collapses the wavefunction. The MANY WORLDS INTERPRETATION instead holds that the universe is cloned at each quantum event, so the moment a cat looks in a box he creates one universe where the scientist is alive and another in which he is dead (*lower opposite*).

Others include the TRANSACTIONAL INTERPRETATION, where the future affects the past, and the BOHM INTERPRETATION, which views the universe as a single entangled whole. More recently, QUANTUM BAYESIANISM (*QBism*), has re-emphasized the role of the observer in the quantum world, suggesting that the wavefunction is not by itself a reflection of an objective reality, but rather a probabilistic representation of an agent's beliefs about the outcome of a measurement.

If photon 1 is measured   *beam splitter*   ...photon 2 can instantaneously
to be spin up...   be shown to be spin down.

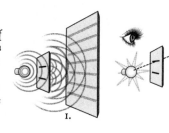

RIGHT: THE TWIN SLIT EXPERIMENT. *Wave or particle?* 1. *Unobserved, a series of single photons, fired at two slits, produce an interference pattern, as if the photons were waves.* 2. *Strangely, on measuring to see which slit each photon passes through, the pattern disappears, suggesting that photons also behave as particles.*

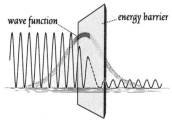

wave function — *energy barrier*

LEFT: QUANTUM TUNNELLING. *An electron meets a solid barrier. However, due to the uncertainty principle, its wave function extends beyond the barrier. There is a small probability of finding it on the other side, and we do. Tunnelling is the basis of electronic components like diodes, transistors, and silicon chips.*

RIGHT: THE CASIMIR EFFECT. *Parallel conducting plates are put in close proximity. Only electromagnetic waves with resonant wavelengths can exist within the cavity, so the field pressure from quantum vacuum fluctuations is greater outside than inside, resulting in the plates being pushed together.*

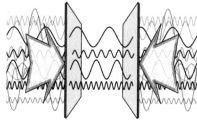

LEFT: SCHRÖDINGER'S BOX. *A subject is trapped in a box with a quantum device that gives a 50/50 chance of survival. The system will remain in a superposition of states, with the hapless subject seemingly both alive and dead, until it is observed. Only when it is measured will the wave function decohere into a single state.*

METAMATERIALS
the future is invisible

Our modern understanding of chemistry enables the manipulation of matter on ever finer scales, producing immensely strong alloys for ocean and space exploration, 3D printers, fuel-cells and solar panels, as well as plastics, polymers, ceramics, and composites that are increasingly flexible, adhesive, superconducting, and heat resistant (*opposite top*).

Many products of this revolution can be found in the microchips and LEDs (*light emitting diodes*) of all manner of electronic gizmos (*below*). SPINTRONICS uses the fundamental properties of matter to allow nuclear magnetic resonance to peek into materials and living tissue, or to read the entangled electrons of quantum computers (*center opposite*).

NANOTECHNOLOGY has the ability to steer atoms on the nanometer scale (10^{-9}m, the length fingernails grow per second), where the intriguing effects of the quantum realm can be harnessed for medicine, electronics, and chemical synthesis (*lower opposite*).

ARTIFICIAL INTELLIGENCE has already deduced millions of novel chemical combinations, and MOLECULAR ENGINEERING will impact further areas of life. Whilst exciting, it is always wise to consider the consequences of using such powerful new toys.

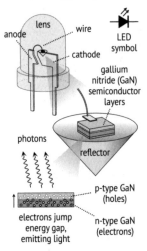

lens — wire
anode
cathode

LED symbol

gallium nitride (GaN) semiconductor layers

photons
reflector

electrons jump energy gap, emitting light

p-type GaN (holes)
n-type GaN (electrons)

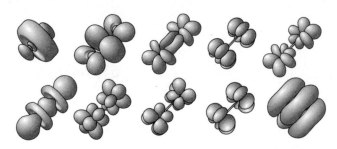

ABOVE: Theoretical QUANTUM CHEMISTRY *predicts the existence of a series of multiply bonded diuranium molecules. They are expected to be stable and have interesting magnetic properties.*

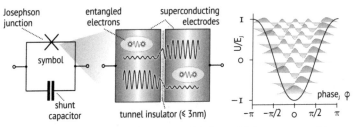

ABOVE: JOSEPHSON JUNCTIONS *allow 'Cooper pairs' of entangled electrons to pass, giving an exact but unequal energy spectrum with a precise frequency to voltage response. They enable SQUIDs (Superconducting QUantum Interference Devices) to measure subtle magnetic fields.*

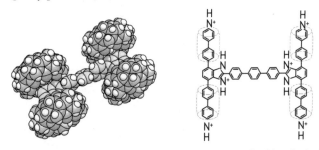

ABOVE: Racing at up to 95nm per hour, and driven by electrons transferred from the tip of an electron microscope, NANOCARS *are built from a hundred or so carbon, hydrogen and other atoms.*

THE PERIODIC TABLE

Group	1	2		3	4	5	6	7	8
	IA	IIA		IIIB	IVB	VB	VIB	VIIB	VIII

Period 1

| H 0 — 1 — **H** — Hydrogen — 1.00079 1310 |

Period 2

| B 4 — 3 — **Li** — Lithium — 6.941 519 | H 5 — 4 — **Be** — Beryllium — 9.01218 900 |

Period 3

| B 12 — 11 — **Na** — Sodium — 22.9878 494 | H 12 — 12 — **Mg** — Magnesium — 24.3050 736 |

Period 4

| B 20 — 19 — **K** — Potassium — 39.0983 418 | C 20 — 20 — **Ca** — Calcium — 40.0785 590 | H 24 — 21 — **Sc** — Scandium — 44.9559 632 | H 26 — 22 — **Ti** — Titanium — 47.867 661 | B 28 — 23 — **V** — Vanadium — 50.9415 648 | B 28 — 24 — **Cr** — Chromium — 51.9961 653 | C 30 — 25 — **Mn** — Manganese — 54.9380 716 | B 28 — 26 — **Fe** — Iron — 55.845 762 |

Period 5

| B 48 — 37 — **Rb** — Rubidium — 85.4678 402 | F 50 — 38 — **Sr** — Strontium — 87.62 548 | H 50 — 39 — **Y** — Yttrium — 88.9059 636 | H 50 — 40 — **Zr** — Zirconium — 91.224 669 | B 52 — 41 — **Nb** — Niobium — 92.9064 653 | B 56 — 42 — **Mo** — Molybdenum — 95.94 694 | H 55 — 43 — **Tc** — Technetium — 97.9072 699 | H 58 — 44 — **Ru** — Ruthenium — 101.07 724 |

Period 6

| B 78 — 55 — **Cs** — Caesium — 132.905 376 | B 82 — 56 — **Ba** — Barium — 137.327 502 | 57 – 70 Lanthanide series ✳ | B 104 — 71 — **Lu** — Lututium — 174.967 531 | H 108 — 72 — **Hf** — Hafnium — 178.49 531 | B 108 — 73 — **Ta** — Tantalum — 180.948 760 | B 110 — 74 — **W** — Tungsten — 183.84 770 | H 112 — 75 — **Re** — Rhenium — 186.207 762 | H 116 — 76 — **Os** — Osmium — 190.23 841 |

Period 7

| H 136 — 87 — **Fr** — Francium — 223.02 381 | H 138 — 88 — **Ra** — Radium — 226.025 510 | 89 – 102 Actinide series ✳✳ | ? 157 — 103 — **Lr** — Lawrencium — 262.110 444 | ? 153 — 104 — **Rf** — Rutherfordium — 263.113 490 | ? 157 — 105 — **Db** — Dubnium — 262.114 ? | ? 157 — 106 — **Sg** — Seaborgium — 266.122 ? | ? 157 — 107 — **Bh** — Bohrium — 264.125 740 | ? 161 — 108 — **Hs** — Hassium — 269.134 ? |

✳ Lanthanide series

| H 82 — 57 — **La** — Lanthanum — 138.906 540 | C 82 — 58 — **Ce** — Cerium — 140.116 665 | H 82 — 59 — **Pr** — Praseodymium — 140.908 556 | H 60 — **Nd** — Neodymium — 144.24 607 | H 84 — 61 — **Pm** — Promethium — 144.913 556 | R 90 — 62 — **Sm** — Samarium — 150.36 540 | B 90 — 63 — **Eu** — Europium — 151.964 548 | B 94 — 64 — **Gd** — Gadolinium — 157.25 594 | H 94 — 65 — **Tb** — Terbium — 158.925 648 |

✳✳ Actinide series

| C 138 — 89 — **Ac** — Actinium — 227.028 669 | C 142 — 90 — **Th** — Thorium — 232.038 674 | T 140 — 91 — **Pa** — Proactinium — 231.0356 568 | ? 146 — 92 — **U** — Uranium — 238.029 385 | O 144 — 93 — **Np** — Neptunium — 237.048 604 | M 150 — 94 — **Pu** — Plutonium — 244.064 585 | H 148 — 95 — **Am** — Americium — 243.061 578 | ? 151 — 96 — **Cm** — Curium — 247.070 581 | C 150 — 97 — **Bk** — Berkelium — 247.070 601 |

IUPAC interim naming system for new elements: o-nil-(n), 1-un-(u), 2-bi-(b), 3-tri-(t), 4-quad-(q), 5-pent-(p), 6-hex-(h), 7-sept-(s), 8-oct-(o), 9-enn-(e)

OF THE ELEMENTS

Crystal	Neutrons	Atomic No.	Symbol	Name	Atomic Weight	First Ionization Energy
	2		He	Helium	4.0026	2370
R	5		B	Boron	10.811	799
H	6		C	Carbon	12.0107	1090
H	7		N	Nitrogen	14.0067	1400
M	8		O	Oxygen	15.9994	1310
M	9		F	Fluorine	18.9984	1680
C	10		Ne	Neon	20.1797	2080
C	14		Al	Aluminium	26.9815	577
C	14		Si	Silicon	28.0855	786
TC	16		P	Phosphorus	30.9738	1060
O	16		S	Sulphur	32.066	1000
O	18		Cl	Chlorine	35.4527	1260
C	22		Ar	Argon	39.948	1520
H	32		Co	Cobalt	58.9332	757
C	30		Ni	Nickel	58.6934	736
C	34		Cu	Copper	63.546	745
H	34		Zn	Zinc	65.39	908
O	38		Ga	Gallium	69.723	577
C	42		Ge	Germanium	72.61	762
R	42		As	Arsenic	74.9216	966
H	46		Se	Selenium	78.96	941
O	44		Br	Bromine	79.904	1140
C	48		Kr	Krypton	83.80	1350
C	58		Rh	Rhodium	102.906	745
C	60		Pd	Palladium	106.42	803
C	60		Ag	Silver	107.868	732
H	66		Cd	Cadmium	112.411	866
T	66		In	Indium	114.818	556
T	70		Sn	Tin	117.710	707
R	70		Sb	Antimony	121.760	833
R	78		Te	Tellurium	127.60	870
C	74		I	Iodine	126.904	1010
C	78		Xe	Xenon	131.29	1170
C	116		Ir	Iridium	192.217	887
C	117		Pt	Platinum	195.078	866
C	118		Au	Gold	196.967	891
R	122		Hg	Mercury	200.59	1010
H	124		Tl	Thallium	204.383	590
C	126		Pb	Lead	207.2	716
M	126		Bi	Bismuth	208.980	703
C	126		Po	Polonium	208.980	812
?	125		At	Astatine	209.987	920
C	136		Rn	Radon	222.017	1040
?	159		Mt	Meitnerium	268.139	?
?	162		Ds	Darmstadtium	272.146	?
?	161		Rg	Roentgenium	272.154	?
?	165		Cn	Copernicium	277	?
?	173		Nh	Nihonium	286	?
?	175		Fl	Flerovium	289	?
?	174		Mc	Moscovium	289	?
?	177		Lv	Livermorium	293	?
?	177		Ts	Tennessine	294	?
?	176		Og	Oganesson	294	?
H	98		Dy	Dysprosium	162.50	657
C	98		Ho	Holmium	164.930	?
H	98		Er	Erbium	167.26	?
H	100		Tm	Thulium	168.934	?
C	104		Yb	Ytterbium	173.04	598
HC	153		Cf	Californium	251.080	608
C	158		Es	Einsteinium	252.083	619
?	157		Fm	Fermium	257.095	627
?	157		Md	Mendelevium	258.098	635
?	157		No	Nobelium	259.101	642

Key to diagram box:

Crystal structure (see below for key) — Number of neutrons (most abundant or stable isotope)

ATOMIC NUMBER

Chemical Symbol

Name of Element

Atomic Weight (average relative mass) — First Ionization Energy (kj mol⁻¹)

B body centred cubic C cubic close packing H hexagonal close packing M monoclinic O orthorhombic R rhombohedral (trigonal) T tetragonal TC triclinic

CONSTANTS & HADRONS

Electron mass	m_e	$9.1091534 \times 10^{-31}$ kg
		$= 0.5110$ MeV
Electron charge	e	1.602189×10^{-19} C
		$= 4.8030 \times 10^{-10}$ esu
Proton mass	m_p	1.672648×10^{-27} kg
		$= 1836.1 \times$ electron mass
Neutron mass	m_n	$1.6749545 \times 10^{-27}$ kg
Atomic mass unit	u	1.66054×10^{-27} kg
Avogadro's no.	N_A	6.022045×1023 mol^{-1}
Bohr radius	a_o	$0.52917706 \times 10^{-10}$ mol^{-1}
Boltzmann const.	k	1.380662×10^{-23} J K^{-1}
Faraday constant	F	9.648456×10^4 C mol^{-1}
Gas constant	R	8.31441 J K^{-1} mol^{-1}
H_2O triple point	T_{tpw}	273.16 K
Ice point temp.	T_{ice}	273.1500 K
Mol. vol. gas	V_m^o	2.241383×10^{-2} m^3 mol^{-1}
Perm. of vacuum	μ_o	$4\pi \times 10^{-7}$ H m^{-1}
Permittivity const.	ε_o	8.8542×10^{-12} F m^{-1}
Planck constant	h	6.626176×10^{23} mol^{-1}
Planck length	l_p	1.616×10^{-35} m
Planck time	t_p	5.319×10^{-44} s
Planck mass	m_p	2.1777×10^{-8} kg
Rcp. fine str. const.	$1/\alpha$	137.036
Rydberg const.	R_H	1.097373×10^7 m^{-1}
Speed of light	c	2.99792458×10^8 m s^{-1}
1 angström (Å)		10×10^{-10} m
1 atmosphere		101325 N m^{-2} (Pa)
1 calorie (cal)		4.184 joules (J)
1 celsius (°C)		273.150 Kelvin (K)
°celsius		$5/9$ (°fahrenheit - 32)
1 curie (Ci)		3.7×10^{10} s^{-1}
1 erg		2.390×10^{-11} kcal
1 esu		3.3356×10^{-10} C
1 eV		1.60218×10^{-19} J
1 eV/molecule		96.485 kJ mol^{-1}
1 kcal mol^{-1}		349.76 cm^{-1}, 0.0433 eV
1 kJ mol^{-1}		83.54 cm^{-1}
1 wave no. (cm^{-1})		2.8591×10^{-3} kcal mol^{-1}

Baryons	Symbol	Mass	Quarks	Charge	Spin
Proton	N$^+$	938	uud	+1	1/2
Neutron	N^o	940	ddu	0	1/2
Sigma$^+$	Σ^+	1198	uus	+1	1/2
Sigma^o	Σ^o	1192	dus	0	1/2
Sigma-	Σ-	1197	dds	-1	1/2
Lambda^o	Λ^o	1116	dus	0	1/2
Xi^o	Ξ^o	1315	uss	0	1/2
Xi$^-$	Ξ^-	1321	dss	-1	1/2
Sigma$^+$	Σ^+	938	uus	+1	1/2
Delta^{++}	Δ^{++}	1231	uuu	+2	3/2
Delta$^+$	Δ^+	1232	duu	+1	3/2
Delta^o	Δ^o	1234	ddu	0	3/2
Delta$^-$	Δ^-	1235	ddd	-1	3/2
Sigma^{*+}	Σ^{*+}	1189	uus	+1	3/2
Sigma^{*o}	Σ^{*o}	1193	dus	0	3/2
Sigma^{*-}	Σ^{*-}	1197	dds	-1	3/2
Xi^{*o}	Ξ^{*o}	1315	uss	0	3/2
Xi^{*-}	Ξ^{*-}	1321	dss	-1	3/2
Omega$^-$	Ω^-	1672	sss	-1	3/2

Mesons					
Pi$^+$	π^+	140	u$\bar{\text{d}}$	+1	0
Pi^o	π^o	135	u$\bar{\text{u}}$, d$\bar{\text{d}}$	0	0
Pi$^-$	π^+	140	d$\bar{\text{u}}$	-1	0
Eta	η^+	547	u$\bar{\text{u}}$, d$\bar{\text{d}}$, s$\bar{\text{s}}$	0	0
Eta prime	η'	958	u$\bar{\text{u}}$, d$\bar{\text{d}}$, s$\bar{\text{s}}$	0	0
Kaon$^+$	K$^+$	494	u$\bar{\text{s}}$	+1	0
Kaon^o	K^o	498	d$\bar{\text{s}}$	0	0
Rho$^+$	ρ^+	770	u$\bar{\text{d}}$	+1	1
Rho^o	ρ^o	770	u$\bar{\text{u}}$, d$\bar{\text{d}}$	0	1
Omega	ω	782	u$\bar{\text{u}}$, d$\bar{\text{d}}$	0	1
Phi	ϕ	1020	s$\bar{\text{s}}$	0	1
K^{*+}	K^{*+}	892	u$\bar{\text{s}}$	+1	1
K^{*o}	K^{*o}	892	d$\bar{\text{s}}$	0	1
J/ψ	ψ	3097	c$\bar{\text{c}}$	0	1
Upsilon	Υ	9460	b$\bar{\text{b}}$	0	1

Note: Over 200 baryons and 36 mesons are known